HOW
CARBON
FOOTPRINTS
WORK

Nick Hunter

W
FRANKLIN WATTS
LONDON•SYDNEY

First published in 2015 by Franklin Watts
338 Euston Road
London NW1 3BH

Franklin Watts Australia
Level 17/207 Kent Street
Sydney, NSW 2000

Produced by Calcium

A CIP catalogue record for this book is available from
the British Library.

ISBN 978 1 4451 3907 4

Dewey classification: 363.7'3874

Printed in China

Franklin Watts is a division of Hachette Children's Books, an Hachette UK company
www.hachette.co.uk

Acknowledgements:
The publisher would like to thank the following for permission to reproduce photographs:
Cover: Shutterstock: Fotofactory tr, Risteski Goce b. Inside: Dreamstime: Devy 2, 10, Djembe
11, Irochka 8, Jlvdream 7, Komelau 9, Maximus117 15, Sabphoto 6, Kirill Solomentsev
18, Wimstime 4, Xneo 14; Shutterstock: Abutyrin 22, Dmitry Berkut 25, Lucian Coman 20,
Sam Cornwel 16, Dalish 23, Dzinnik Darius 5, Marcio Eugenio 17, Daria Filimonova 27,
Innershadows Photography 24, Ixpert 26, Kurhan 13, Bartlomiej Magierowski 21, Photobank.
ch 1, 29, Dasha Rosato 12, Fedor Selivanov 19, Ssuaphotos 28.

Every attempt has been made to clear copyright. Should there be any inadvertent omission
please apply to the publisher for rectification.

Contents

Your carbon footprint

If you walk along a beach or a muddy path, you leave a footprint. Footprints in the sand are not the only way in which people leave a mark on their surroundings – we also leave behind an invisible footprint with almost everything that we do. This is our carbon footprint.

What is a carbon footprint?

A carbon footprint is the measurement of how much carbon dioxide and other greenhouse gases (GHGs) are released into the air as a result of our actions. For example, carbon dioxide is released by power stations burning coal and gas to generate electricity, so every time we switch on a light, we cause more carbon to be released.

We cannot see our carbon footprints but we can measure their effects over long periods of time.

GHGs up close

Your carbon footprint is not just about carbon dioxide. Other GHGs also cause climate change and these are included in your carbon footprint. Another term for carbon footprint is 'carbon dioxide equivalent (CO2e)'. Other GHGs include:

- Methane (from farming and landfill): 25 times more harmful than carbon dioxide.
- Nitrous oxide (from industry): 1,000 times more harmful than carbon dioxide.
- Gases, such as sulphur dioxide and propane, (in refrigeration): 1,000 times more harmful than carbon dioxide.

Why does it matter?

The size of your carbon footprint matters because the combined footprint of all the people and animals on Earth changes the mixture of gases in the atmosphere. We rely on this mix of gases for the air we breathe, to protect us from the rays of the sun and to provide the weather conditions that make life possible. There is clear evidence that the carbon footprint of 7 billion human beings on Earth is changing this delicate mix of gases. The GHGs trap heat from the sun and are causing Earth's climate to get warmer.

The food we eat depends on Earth's climate. If it is too hot or too cold, nothing grows.

Carbon everywhere

Carbon is one of the most important substances on Earth. It makes up less than 0.1 per cent of Earth's crust, but it is an essential part of all plants and animals. Carbon is found in the food we grow and eat, as well as in the fuels, such as coal and oil, we burn to give us warmth and to power our industries.

Oxygen from the air reacts with our food to release energy. We breathe out carbon dioxide that is produced in this chemical reaction.

Releasing carbon

Animals, including humans, release carbon into the atmosphere in the form of carbon dioxide every time they breathe. Carbon dioxide is also released when materials such as wood, oil and coal are burnt. Oil and coal are called fossil fuels because they formed from the remains of prehistoric plants and animals over millions of years. During this time, the carbon inside them was trapped. Burning these fossil fuels releases the carbon into the atmosphere.

Absorbing carbon

With all these ways of releasing carbon dioxide, why does the atmosphere not fill up with carbon? The answer lies in the world's forests. Plants absorb carbon dioxide through a process called photosynthesis. Huge areas of forest, such as the vast Amazon rainforest in South America, act like a sponge and soak up carbon dioxide.

The Amazon basin of South America is the largest area of tropical rainforest in the world.

Carbon emissions up close

There are two main parts to your own carbon footprint:

- Direct carbon emissions: if you use a plastic cup, carbon was directly released in making the cup and shipping it to you, possibly from the other side of the world.
- Indirect carbon emissions: these are caused by all the other people and processes that helped the cup to reach you. They could include mining for the oil that was used to make the plastic in the cup, or the emissions from the luxury car of the person who owns the factory where the cup was made.

The trouble with carbon

If carbon is part of all living things, and plants and trees remove it from the atmosphere, why should we worry about our carbon footprint? To answer this question we need to understand the effects of carbon dioxide and other GHGs in the atmosphere.

Coral reefs are home to a huge variety of living things. They are under threat from warming seas and also because absorbed carbon dioxide is changing the oceans.

The growth of industry

For most of human history, there was no problem with carbon. The trouble started in the 1700s when factories began to burn coal. At the same time, forests were being cut down to make way for farmland and cities. The levels of carbon dioxide in the atmosphere began to increase.

A growing problem

Things got worse in the 1900s, as electricity became the main source of power for homes and factories. Cars, aircraft and other forms of transport also then developed, and these all used oil-based fuels. Carbon emissions around the world rocketed.

More GHGs

The increase in carbon emissions has led to a sharp increase in the quantity of GHGs in the atmosphere. These gases trap heat that radiates from Earth's surface, stopping it from being lost in space. As more heat is trapped, the atmosphere gets warmer. This is known as 'the greenhouse effect'.

Even small changes in temperature can make a big difference. It can cause large areas of ice in the Arctic and Antarctic to melt. In hotter parts of the world, droughts could be more severe. This would prevent millions of people from growing the food that they need.

If harvests fail because of drought, millions of people will go hungry.

ECO FACT

Temperatures rising

In 1957, there were 320 parts of carbon dioxide per million in the atmosphere. In just over 50 years, the level of carbon dioxide has risen to almost 400 parts per million. Over the last century, the average surface temperature on Earth has risen by around 0.7 °C.

The world's carbon footprint

Scientists say swift action is needed to reduce the carbon footprint of Earth's population and slow the progress of global climate change. But how big a problem is it, and is everyone equally to blame for the rising carbon emissions?

How much carbon?

Human activities produce an astonishing 34.5 billion tonnes of carbon dioxide every year. Although we know all about the risks of climate change, carbon emissions are still increasing. Carbon emissions have grown by almost half since 1992.

Of course, we do not all have the same-sized carbon footprint. If you are reading this book in a developed country like the United Kingdom, you probably account for more than 3.6 tonnes of carbon dioxide per year. If you live in Africa, however, you probably have a much smaller carbon footprint.

Aircraft are one of the fastest-growing causes of carbon emissions, with air travel expected to grow by more than 30 per cent between 2012 and 2017.

ECO FACT

Invisible emissions

The world's carbon emissions are measured in billions of tonnes. The best way to understand it is with specific examples, especially as we cannot see GHGs. For example, a powerful car driven for 1,600 kilometres would release around 0.4 tonnes of carbon dioxide.

The biggest footprints

China produces more carbon dioxide than any other country, and was responsible for 7.5 billion tonnes in 2010. However, China is also home to more people than any other country. China produced 5.4 tonnes of carbon dioxide per person in 2012, which is well behind the 16.3 tonnes per person produced in the United States and 18.8 tonnes in Australia. In contrast, many of the world's poorest countries produce less than 1 tonne of carbon dioxide per person. These figures include only direct carbon emissions and some estimates are much higher.

Calculating your carbon

Your own carbon footprint depends on where you live, the transport you use, how careful you are to limit your electricity use and even the type of food you eat. One way to calculate your carbon footprint is to keep a diary of everything you do that produces carbon.

The carbon culprits

Online carbon calculators help you to get an idea of your carbon footprint. However, they often only consider actions such as direct energy use and transport, without noting the other ways you produce carbon (see page 7).

If you can travel by bicycle rather than in a car, that will help to reduce your carbon footprint.

How to reduce your carbon footprint

These are the main areas to think about when trying to reduce your carbon footprint.

- In the United Kingdom, more than 15 per cent of carbon dioxide is produced by businesses, including the manufacture of the things we buy. The more things you throw away after one use, such as plastic water bottles, the bigger your carbon footprint.
- Australia produces about 5 per cent of its carbon dioxide through manufacturing and industry.
- Personal travel is a big area, with every car or aircraft journey pumping more carbon dioxide into the atmosphere.
- Energy we use at home is another big part of our carbon footprint, particularly heating and washing.
- Farming and food transportation: has your food been transported from across the world or is it grown near your home, which produces fewer carbon emissions?

It is hard to think of anything you do that does not cause carbon emissions. However, that also means there are a lot of ways you can reduce your carbon footprint.

In the United Kingdom, people drink around 2 billion litres of bottled water a year. This produces 350,000 tonnes of carbon dioxide and a pile of plastic that takes up to 400 years to decompose.

Farms up close

Some foods produce more GHGs than others. Meat produces more carbon than vegetables and cereal crops because animals need energy to live and move around on a farm. If you ate a burger every day, that would produce around 1 tonne of carbon dioxide in a year. It would not be very good for your body either!

Countering carbon

Our biggest allies in the fight to reduce carbon footprints are the trees and plants that absorb carbon. Preserving Earth's precious forests, particularly in tropical regions, is just as important as reducing our own carbon footprints.

Wildfires release millions of tonnes of carbon dioxide, as well as endangering lives and properties.

Rainforests

Tropical rainforests are found in areas close to the equator. They include the Amazon rainforest in South America and other rainforests in Central America, Africa and Southeast Asia. Tropical rainforests cover about 7 per cent of Earth's land area today, but in 1800 they covered twice as much land. Much of these precious forests has been lost to make way for farmland and other human uses. The tropical rainforests contain almost half of all the carbon in plants around the world.

Burning rainforests

When forests are burnt, the carbon they contain is released in the form of carbon dioxide, and the area of forest that can absorb carbon is reduced. Preserving this forest is essential to minimise the impact of the human carbon footprint. We can help to preserve rainforests by not buying products made from rainforest trees, such as furniture that is made from mahogany and teak.

Fuels from crops

Biofuels are produced from crops and can be used to power vehicles. Because the plants that make biofuels absorb carbon dioxide while they grow, they have a lower carbon footprint than fossil fuels. However, to help the crops grow, fertilisers are used. This releases GHGs into the atmosphere. Forests are also sometimes cleared in order to create areas in which biofuel crops can be grown.

More and more land is being used to grow biofuel crops, but this means there is less space to grow essential food crops.

ECO FACT

Tree trouble

Trees may not always be the best solution to climate change. In snowy regions, heat from the sun is reflected back into space by the white snow. Planting more trees in these areas could actually increase the pace of climate change, even though the trees absorb carbon from the atmosphere. This is because the trees reduce the snowy areas that reflect the sun's rays back into space.

At home and at work

Take a look around your home and see if you can find areas where energy is wasted, such as lights or appliances left on in empty rooms. If you can find ways to reduce your family's carbon footprint, you will probably save some money, too.

Heating and cooling

Most carbon emissions at home are produced by heating rooms and heating water for washing. Heat often escapes through roofs, walls and draughty windows and doors.

Energy-efficient lightbulbs can reduce our carbon footprint, but not if we leave them on when we do not need them.

Insulation for roof spaces and walls is made from materials that do not allow heat to pass through them easily. Double-glazed windows are also important in preventing heat loss. Air conditioning in workplace buildings uses just as much energy as heating but there are other ways to keep buildings cool. Tinted or reflective double glazing prevents heat from getting in. Reflective blinds on windows also help to keep out the hot sun.

Computers and other electrical appliances add to carbon footprints, but heating and air-conditioning systems have a greater impact.

Electrical appliances

Watching TV for one hour a day produces the same amount of carbon dioxide in a year as a 45-minute journey by car. If you leave the TV on standby the rest of the time, it will actually produce almost as much carbon in a year with no one watching it. If there is a light glowing on the front of the device, then your TV, computer or any other device is adding to your footprint.

ECO FACT

Brilliant bulbs

Almost 95 per cent of the energy used by an old-fashioned lightbulb is wasted as heat energy. Energy-efficient bulbs release less heat so they use less energy and have a smaller carbon footprint.

On the move

Transport is one of the main areas where individual choices can make a big difference. Transport is now the fastest-growing source of GHG emissions. About one-quarter of the United Kingdom's carbon footprint comes from travel, and it is responsible for about one-eighth of Australia's.

Car footprint

Almost all of the 1 billion cars on the roads around the world burn petrol to provide them with energy. The carbon footprint of your car depends on the size of its engine and the types of journey you make. You could reduce its carbon footprint by making fewer journeys. Could you walk or bicycle to school?

Many people enjoy driving a large car, but these vehicles use up more petrol than smaller cars.

Technology for plug-in hybrid cars is improving all the time.

These cars use a petrol engine only when the charge on their battery runs low.

New cars needed

If the world's carbon footprint is to reach manageable levels, new cars are needed. Hybrid cars use an electric motor and a conventional engine to reduce carbon emissions. In the future, most cars could be powered by electricity or clean fuels, such as hydrogen, that produce no carbon dioxide.

Air travel

If you travel by air, even if it is just once a year, this will make up a big part of your carbon footprint. Aircraft use much more fuel per kilometre than cars, and people travel further by plane than they would in a car. A round-trip between the United Kingdom and North America produces 3.6 tonnes of carbon for each passenger, which is roughly the same as the average car produces in a year.

Cars up close

Here are some tips to reduce the carbon footprint of a car:

- Take as many passengers as possible. The more people there are in the car, the smaller each person's footprint will be.
- Use a smaller car, a hybrid car or public transport if you can.
- Try not to travel at busy times. Starting and stopping in busy traffic can use three times as much energy as driving smoothly.

Low-carbon shopping

It is easy to see how burning fuel in cars or heating our homes produces carbon dioxide, but what is wrong with shopping? To understand how it affects our carbon footprint, we need to look far beyond the shops.

From factory to shop

The biggest contribution to the world's carbon footprint is industry. China, which produces more GHGs than any other country, is also home to the factories that make so many of the consumer goods and clothes you see in your local shops. It takes a lot of energy to get metals and other materials out of the ground and refine them for use. Plastics are all made from oil. Manufactured goods also have to be transported across the world for us to buy them.

If more people buy local goods or products with less packaging, this will persuade supermarkets to stock more of these goods.

The 3Rs

There are three main ways to reduce your carbon footprint when you shop:

- Reduce – think about what you need and what you do not need before buying things. Look for goods that have less packaging, as this can use as much energy as the goods themselves.
- Reuse – never buy anything that you use only once. For example, a water bottle can be refilled and used many times.
- Recycle – recycling generally uses less energy than creating entirely new goods. It also cuts down on waste.

Producing goods in China and other countries helps to reduce prices, but there is a price to pay in the carbon footprint of transporting those goods.

ECO FACT

Food miles

Transporting food – particularly perishable food, such as fruit and vegetables – produces carbon emissions. Food is often transported by air to keep it fresh. Strawberries that are flown in from overseas will have 10 times the carbon emissions of any strawberries grown locally and in season.

Carbon killers

There are many ways that we can each reduce our carbon footprint. However, we can do absolutely nothing about some of the biggest sources of GHGs.

Natural causes

GHGs are released when anything burns, and some of the biggest explosions on Earth are beyond our control. An erupting volcano throws out many millions of tonnes of carbon. However, it would be misleading to think of volcanoes as contributing to climate change. Big eruptions can change weather over short periods, but the carbon produced by volcanoes is less than 1 per cent of the carbon produced by human activities.

Coal has a bigger carbon footprint than other fuel. Despite this, coal use is still rising in countries such as China, the United Kingdom and Australia.

This geothermal power station in Iceland heats homes, offices and even a swimming pool.

ECO FACT

High-altitude emissions

We have already seen that air travel is a big contributor to the world's carbon footprint. The full story is even more alarming. Fumes from aircraft produced high in the atmosphere actually have a greater effect on climate than the fumes released at ground level.

Power problem

The world's carbon footprint could be cut dramatically if countries and power-generation companies changed the fuels that they use to generate electricity. Around three-quarters of electricity in the United Kingdom is generated by burning fossil fuels, and just over one-third in Australia. Coal is still the number one fuel used by power stations around the world.

Cleaner energy

Carbon emissions could be reduced by using more natural gas or, better still, renewable energies, such as solar and wind power. Iceland has abundant supplies of power from hydroelectric dams and geothermal energy from underground, meaning that people can use as much power as they want, whenever they need it, without causing additional and harmful carbon emissions.

Making a difference?

The world's carbon footprint is so vast that our own part in it can seem insignificant. After all, what is the point of reducing your carbon footprint if more than 1 billion people in China are increasing their carbon dioxide emissions?

If you have space to grow food, this can be rewarding and healthy, as well as reducing the cost and carbon footprint of your food.

Influencing others

If you decide to reduce your own carbon footprint, you can also convince your friends and other people to do the same. You can find ways to tell people in your school and community about their carbon footprints and suggest steps they can take to cut their impact. You could even suggest ways to reduce your school's carbon footprint.

Changing others

Just as importantly, your actions can change the way that corporations and governments behave. If corporations see people choosing to buy low-carbon products or driving hybrid cars, they will put more time and money into these products.

Politicians and pressure groups

Changes to reduce the carbon footprint of cars and power stations often happen because of government laws. You can write to local or national politicians about taking steps to reduce carbon emissions. There are also many non-governmental organisations and pressure groups that work to tackle carbon emissions and inform people about the issue.

ECO FACT

Some people feel so strongly about climate change and pollution that they join pressure groups to make their views heard.

Predicting the future

Scientists are in no doubt that climate change is happening, but predicting how much carbon we will produce in the future and what effect it will have is much more difficult. The Intergovernmental Panel on Climate Change has predicted that global temperatures could rise by anywhere between 1.1 and 6.4°C by 2099.

Carbon trading and offsetting

Carbon trading is one way that the world's governments have tried to reduce carbon emissions. It means that businesses now have to pay for the GHGs they produce. Businesses and individuals can also offset, or make up for, their own GHG emissions.

Climate change is a global problem but countries have often argued about how they should work together to solve it.

Trading carbon

Carbon trading works by giving industries a limited amount of carbon that they are allowed to produce. If they want to increase their emissions, they have to buy credits from another business that has not used all its credits. This rewards industries and countries that take steps to try to reduce their carbon emissions.

Offsetting

Many businesses try to reduce their carbon footprints by offsetting the carbon they produce. Ways to do this include paying for trees to be planted to help to absorb carbon dioxide, and putting money into renewable energy – such as by distributing energy-saving lightbulbs in developing countries.

Tree planting

Some businesses, such as airlines, will give customers the option to pay towards tree planting to offset the extra carbon emissions of their flight. While carbon offsetting may well be better than no action at all, critics say that it does not really tackle the main issue of developing low-carbon lifestyles and technology.

Planting a tree improves your local environment and may help to offset your carbon footprint.

Carbon offsetting up close

Carbon offsetting may make people feel better about choosing to take a long-haul flight, but does it really make any difference? Here are two problems with offsetting:

- It may take a very long time for trees that are planted to absorb the carbon dioxide that is being produced.
- If we are to reduce our carbon footprint, offsetting should happen in addition to reducing carbon emissions, not instead of it.

27

Footprints in the future

We now understand more than ever before about the invisible footprint that we leave on the planet. However, we are still a long way from slowing the pace of climate change. The race is on to find new ways of reducing the world's carbon footprint before it is too late.

Future power

Power stations are still heavily reliant on fossil fuels. Wind, wave and solar energy are all ways of generating electricity with almost no carbon footprint, but these sources currently provide only a very small amount of our energy. Solutions to this problem include building giant solar power stations to be powered by the constant sunshine in desert regions, or making solar panels part of every new building.

For wind energy to provide more than a small part of our energy needs, huge areas would need to be covered with wind turbines.

New technology

Scientists are also working on technology to capture the carbon produced by our fossil-fuel power stations and store it underground so it does not reach the atmosphere.

Carbon-free cars

New materials will also help to reduce the carbon footprint of the average car. A car uses much of its energy to push itself, rather than the people or luggage it carries. Scientists are working at a microscopic level to create ultra-light but strong materials for the next generation of cars. These cars will be powered by plug-in electric motors or even by hydrogen fuel cells, producing no carbon emissions.

New buildings can use high-tech or natural materials to provide insulation and reduce their carbon emissions.

ECO FACT

Future challenges

Earth's population is expected to rise from the 7 billion people on the planet today to 9 billion people by 2050. The challenge, with a more crowded world, is to keep our carbon footprint at the same level as today, or even to lower it.

Glossary

atmosphere a layer of gases surrounding Earth and containing the oxygen that humans and other animals breathe

carbon dioxide (CO_2) a GHG that is released when fossil fuels and organic matter are burnt. Carbon dioxide is also naturally present in Earth's atmosphere and is absorbed by plants

carbon footprint the amount of GHG emissions caused by a single person or organisation

carbon trading a system by which businesses have a fixed allowance of carbon emissions that they can trade with others. For example, if an organisation does not use up all of its carbon allowance, it can sell the remaining carbon allowance to another organisation that has used its entire allowance

conventional normal or usual, such as conventional cars powered by an internal combustion engine

developing countries countries where most people have a low standard of living, little modern industry and an economy that is still developing

fossil fuels energy sources, including coal, oil and natural gas, formed from the decayed remains of living things

generate to convert one form of energy into another, for example to produce electricity

geothermal heat and energy from beneath Earth's crust

greenhouse gases (GHGs) gases that absorb heat in the atmosphere

hybrid cars cars that have two sources of power, such as an electric motor and a conventional petrol engine

hydrogen a flammable gas that combines with oxygen to make water

insulation material that does not conduct heat and can prevent heat energy from escaping, such as from a building

petrol a flammable product from crude oil that is used in internal combustion engines

refrigeration the process of keeping things cool, such as food

renewable material that can be renewed once it is used, such as plants that can be replaced. Fossil fuels, which take millions of years to form, are not renewable

solar from the sun

For more information

Books

Climate Change (Eyewitness), John Woodward,
Dorling Kindersley

How Bad Are Bananas?: The carbon footprint of everything,
Mike Berners-Lee, Profile Books

*How to Reduce Your Carbon Footprint: 365 Practical Ways to
Make a Difference*, Joanna Yarrow, Duncan Baird Publishers

Reducing the Carbon Footprint, Anne Rooney, Franklin Watts

Websites

Find out more about carbon footprints, global warming and
a lot of other environmental issues at:
**http://news.bbc.co.uk/cbbcnews/hi/newsid_6720000/
newsid_6720600/6720673.stm**

This carbon-footprint calculator works out your carbon footprint:
www.nature.org/greenliving/carboncalculator

Discover loads of information about how to be friendly to
our planet at:
www.earthfirst.net.au/footprint-calculators.html

Note to parents and teachers
Every effort has been made by the Publisher to ensure that these websites contain
no inappropriate or offensive material. However, because of the nature of the
Internet, it is impossible to guarantee that the contents of these sites will not be
altered. We strongly advise that Internet access is supervised by a responsible adult.

Index